Viewing Life Mathematically

+ Integrated Review

Guided Notes

A division of Quant Systems, Inc.
546 Long Point Road, Mount Pleasant, SC 29464

Printed in the United States of America

Table of Contents

Reading and Writing Whole Numbers

o   The symbols used in our number system are called _____.

o   The value of a digit depends on its position to the left of a beginning point, called a _____ point.

o   Fill in the missing items in the table below.

1,000,000,000	Billions
100,000,000	
10,000,000	Ten millions
	Millions
100,000	Hundred thousands
10,000	
	Thousands
100	Hundreds
10	
1	Ones

o The _____ numbers are the number 0 and the natural numbers.

o _____ notation can help you to translate a whole number into its English word equivalent.

o Example:

7,034,165 = 7,000,000 + 30,000 + 4000 + 100 + 60 + 5

7,034,165 is read as

Seven _____, _____ thousand, one hundred sixty-five

Rounding and Estimating with Whole Numbers

- To _____ a given number means to find another number close to the given number. The desired place of accuracy must be stated.

- Rounding Rule for Whole Numbers

  1. Look at the single digit just to the _____ of the digit that is in the place of desired accuracy.

  2. If this digit is _____ or greater, make the digit in the desired place of accuracy one larger and replace all digits to the _____ with zeros. All digits to the left remain unchanged unless a 9 is made one larger; then the next digit to the left is increased by 1.

  3. If this digit is less than 5, leave the digit that is in the place of desired accuracy as it is, and replace all digits to the right with _____. All digits to the left remain _____.

- One use for rounded numbers is to _____ an answer before any calculations are made with the given numbers.

- Example:

Estimate the **sum** of 83 and 19 using rounded values, then find the sum.

Estimated sum: _____   Actual sum: _____

o Example:

Estimate the **product** of 83 and 19 using rounded values, then find the product.

Estimated product: _____   Actual product: _____

Exponents and Order of Operations

o When looking at 5^3, you have the following parts:

  ▪ The base is: _____

  ▪ The exponent is: _____

  ▪ The product is: _____

  ▪ The exponential expression is: ___ ___ _____

o In expressions with exponent 2, the base is said to be _____.

o In expressions with exponent 3, the base is said to be _____.

o With other exponents, the base is said to be "_____."

o For any number, a, $a^1 = a$. An example would be $7^1 =$ _____.

o For any nonzero number, a, $a^0 = 0$. An example would be $7^0 =$ _____.

o The rules for order of operations are:

  1. Simplify within _____ _____, such as parentheses
      ( ), brackets [ ], or braces { }. Start with the _____ group.

  2. Evaluate any numbers or expressions indicated by _____.

  3. Moving from _____ to _____, perform any _____
      or _____ in the order in which it appears.

  4. Moving from _____ to _____, perform any _____
      or _____ in the order in which it appears.

Problem Solving with Whole Numbers

o   Basic Strategy for Solving Word Problems:

1.   _____ the problem carefully.

2.   _____ any type of figure or _____ that might be helpful

and decide what operations are needed.

3.   Perform the _____ to solve the problem.

4.   _____ your work.

o   Key words when solving word problems are:

Addition	Subtraction	Multiplication	Division

o   To find the average of a set of numbers, you will need to:

1. Find the _____ of the given set of numbers.

2. _____ this sum by the total number of numbers in the set. This

    quotient is called the _____ of the given set of numbers.

Translating English Phrases and Algebraic Expressions

Addition	Subtraction	Multiplication	Division	Exponent (Powers)
Add	Subtract (from)	Multiply	Divide	Square of
Sum	Difference	Product	Quotient	Cube of
Plus	Minus	Times		
More than	Less than	Twice		
Increased by	Decreased by	Of (with fractions and percents)		
	less			

Key Words for Translating Phrases

- Remember that in each case "a number" or "the number" implies the use of a _____ (an unknown quantity).

- Be very careful when writing and/or interpreting expressions indicating _____. Be sure that the subtraction is in the order indicated by the wording in the problem. The same is true with expressions involving_____.

- _____ and _____ are done with the values in the same order that they are given in the problem.

Solving Linear Equations: *ax+b=c*

o The general procedure for solving linear equations is now a _____ of the procedures stated in another section.

o Steps for solving equations in the *ax+b=c* format.

1. _____ like terms on both sides of the equation.

2. Use the _____ principle of equality and add the opposite of the constant *b* to both sides.

3. Use the _____ (or division) principle of equality to multiply both sides by the reciprocal of the coefficient of the variable (or divide both sides by the coefficient itself). The coefficient of the variable will become +1.

4. Check your answer by _____ it for the variable in the original equation.

o Keep in mind that checking can be quite _____-_____ and need not be done for every problem. This is particularly important on _____. You should check only if you have time after the entire exam is _____.

Example:

$3x - 8 = 28$ Add 8 to each side.

$3x = 36$ Divide both sides by 3.

$x = 12$

The Real Number Line and Inequalities

o A number line is a picture of different types of _____ and their

   relationships to each other.

o The graph of a number is the point that _____ to the number and the

   number is called the _____ of the point.

o On a horizontal number line, the point one unit to the left of 0 is

   the _____ of 1. It is called negative 1 and is symbolized -1.

o The negative sign $(-)$ indicates the _____ of a number as well as a

   _____ number. It is also used to indicate _____.

o The set of numbers consisting of the _____ numbers and their

   _____ is called the set of integers: $\mathbb{Z} = \{..., -3, -2, -1, 0, 1, 2, 3, ...\}$.

o The _____ numbers are also called _____ integers.

   Their opposites are called _____ integers. Zero is its

   _____ opposite and is neither positive nor negative $(0 = -0)$.

o The opposite of a positive integer is a _____ integer, and the opposite of a

   negative integer is a _____ integer.

o A rational number is a number that can be written in the form of $\frac{a}{b}$, where a and b are

   _____ and $b \neq$ _____. ($\neq$ is read "is not equal to").

o A rational number is a number that can be written in _____ form as a

   _____ decimal or as an infinite _____ decimal.

o Irrational numbers can be written as infinite _____ decimal numbers.

o   All rational numbers and irrational numbers are classified as _____

numbers (ℝ) and can be written in some decimal form.

o   Symbols of inequalities:

o   = is _____ to

o   ≠ is _____ equal to

o   < is _____ than

o   > is _____ than

o   ≤ is less than or _____ to

o   ≥ is greater than or _____ to

Addition with Real Numbers

o   The sum of two positive real numbers is _____.

o   The sum of two negative real numbers is _____.

o   The sum of a positive real number and a negative real number may be

_____ or _____ (or _____)

depending on which number is _____ from 0.

o   To add two real numbers with like signs:

1.   Add their _____ _____

2.   Use the _____ sign

o   To add two real numbers with unlike signs:

1.   _____ their absolute values (the smaller from the larger)

2.   Use the sign of the number with the _____ absolute value

o   The _____ sign may be omitted when writing _____

numbers, but the _____ sign must always be written for

_____ numbers.

o   If there is no sign in front of a real number, the real number is understood to be

_____.

o   A number is said to be a _____ or to _____ an

equation if it gives a true statement when substituted for the variable.

Subtraction with Real Numbers

o The opposite of a real number is called its _____

_____.

o The sum of a number and its additive inverse is _____.

o Adding positive numbers "_____ _____" or "_____"

the numbers in a _____ direction (moving _____ on the

number line)

o Adding negative numbers "_____ _____" or

"_____" the numbers in a _____ direction (moving

_____ on the number line).

o In subtraction, you want to find the "_____" two

numbers.

o To subtract, _____ the _____ of the number being

subtracted.

o To find the change in value between two numbers, take the end value and

_____ the beginning value.

o This process looks like:

Change in value = (_____ value) − (_____ value)

Multiplication and Division with Real Numbers

o   If a and b are positive real numbers, then:

     o   The product of two positive numbers is _____: $a \cdot b = +ab$.

     Example: $8(14) = 122$

     o   The product of two negative numbers is _____: $(-a)(-b) = +ab$.

     Example: $-4(-5) = 20$

     o   The product of a positive number and a negative number is _____: $a(-b) = -ab$.

     Example: $10(-3) = -30$

     o   The product of 0 and any number is _____: $a \cdot 0 = 0$ and $(-a) \cdot 0 = 0$.

     Example: $71(0) = 0$

o   If a and b are positive real numbers (where $b \neq 0$), then:

     o   The quotient of two positive numbers is _____: $\dfrac{a}{b} = +\dfrac{a}{b}$

     Example: $\dfrac{14}{2} = 7$

     o   The quotient of two negative numbers is _____: $\dfrac{-a}{-b} = +\dfrac{a}{b}$

     Example: $\dfrac{-18}{-6} = 3$

     o   The quotient of a positive number and a negative number is _____:

$\dfrac{-a}{b} = -\dfrac{a}{b}$ and $\dfrac{a}{-b} = -\dfrac{a}{b}$

     Example: $\dfrac{-144}{12} = -12$

     Example: $\dfrac{88}{-11} = -8$

o   Quick Reference Facts with Multiplication and Division with Real Numbers:

   o   If the numbers have the _____ sign, both the product and quotient will be

        _____.

   o   If the numbers have _____ signs, both the product and quotient will be

        _____.

Order of Operations with Real Numbers

o   Order of Operations:

     o   Simplify within grouping symbols, such as

        _____ ( ), _____ [ ], and

        _____ { }, working from the innermost grouping

        _____.

     o   Find any powers indicated by _____.

     o   Moving from _____ to _____, perform any

        _____ or _____ in the order they appear.

     o   Moving from _____ to _____, perform any

        _____ or _____ in the order they appear.

o   Keep in mind that there are other _____ symbols like the

  _____ _____ bars (such as $|3+5|$), the

  _____ _____ (as in $\dfrac{14+7}{3}$), and the

  _____ _____ symbol (such as $\sqrt{40+9}$).

o   Even though the mnemonic PEMDAS is helpful, remember that multiplication

and division are performed as they _____, left to right.

o   Also, addition and subtraction are performed as they _____, left to right.

Introduction to Fractions and Mixed Numbers

o   The parts of a fraction are (shown below):

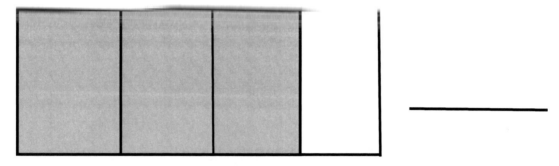

o   Fractions are used to indicate _____ of a whole.

o   Use the picture below to write a fraction representing the shaded portion of the shape:

_____

o   The fraction $\frac{2}{7}$ represents _____ of _____ equal parts.

o   Whole numbers can be thought of as fractions with a denominator of

_____.

o   Fraction notation represents the operation of _____.

o   There are two rules to keep in mind when working with the value of zero in a fraction:

o   For any nonzero value, b, $\frac{0}{b} = 0$. An example of this would be $\frac{0}{4} = 0$.

o   For any value of a, $\frac{a}{0} =$ undefined. An example of this would be $\frac{6}{0} =$ undefined.

o   Steps to multiply fractions:

   1. Multiply the _____.

   2. Multiply the _____.

o Multiply the following fraction: $\dfrac{2}{3} \cdot \dfrac{1}{5} =$ _____ = _____

o To find an equivalent fraction, you need to _____ the numerator and the

   denominator by the _____ nonzero whole number.

o Find an equivalent fraction for $\dfrac{2}{3} =$ _____ by multiplying both the numerator and the denominator

   by 5.

o There are two steps to reducing fractions to lowest terms:

   1. Factor the _____ and _____ into prime factors.

   2. Use the fact that $\dfrac{k}{k} = 1$ and divide out all of the _____ factors.

o What is the common factor that can be divided out of the fraction, $\dfrac{4}{12}$?

o What is $\dfrac{4}{12}$ in lowest terms?

o A mixed number is the sum of a _____ and a _____

   fraction.

o To change a mixed number to an improper fraction, you need to:

   1. Multiply the whole number by the _____ of the proper fraction.

   2. Add the _____ of the proper fraction to this product.

   3. Write this _____ over the denominator of the fraction.

o Change $3\dfrac{1}{2}$ to an improper fraction:

   Multiply the whole number by the denominator: _____ · _____ = _____

   o Add the numerator to the product from above: _____ + _____ = _____

   o Write this sum over the denominator: _____

Introduction to Decimal Numbers

o   Decimal notation uses a _____ _____ system and a

_____ point, with whole numbers written to the

_____ and fractions written to the _____ of the

decimal point.

o   To read or write a decimal number:

    o   _____ (or write) the whole number.

    o   Read (or write) the word "_____" in place of the decimal point.

    o   Read (or write) the _____ part as a whole number. Then name the fraction

    with the name of the last _____ on the right. Add "th" to the end of the

    fraction place value.

    o   Remember that if there is not a whole number, you can put a _____ to the left

    of the decimal point.

o   Write the following mixed number as a decimal number:

$$3\frac{4}{100} = \underline{\quad}.\underline{\quad}\ \underline{\quad}$$

    and it would be read as _____ AND _____ _____

o   When writing seventeen and 5 thousandths in decimal notation you would have

_____ holders in the tenths and hundredths place values.

It would look like 17.005.

o   When comparing decimals:

    o   Moving _____ to _____, compare digits with the same

    _____ value.

    o   When one compared digit is larger, the _____ is larger.

o   Compare the following values:

5.789   Notice that the numbers are lined up, for easier comparison.
5.754

When moving from left to right, the digits are the same until the _____ place

values. Those are going to be used to compare. Since the 8 is a larger value than 5,

_____ is a larger value.

o   Rules for rounding decimals:

   o   Look at the digit to the _____ of the place of desired accuracy.

   o   If this digit is 5 or _____, make the digit in the desired place of accuracy one

   larger and replace all digits to the right with zeros.

   All digits to the left remain unchanged unless a 9 is made one larger. This effectively changes

   the 9 to 10 which means the next digit to the left must be increased by 1.

   o   If this digit is _____ than 5, leave the digit in the desired place of accuracy as

   it is and replace all digits to the right with zeros.

   All digits to the left remain unchanged.

   o   Zeros **to the right of the place of accuracy** and to the right of the decimal point must be

   _____. In this way the place of accuracy is clearly understood. If a rounded

   number has a 0 in the desired place of accuracy, then that 0 remains.

Round 13.2687 to the nearest hundredth.

   The digit in the hundredths is _____ .

   The digit in the place value to the right of the hundredths is _____ .

   Since that digit is greater than 5, the 6 in the hundredths place value changes to a 7.

   The rounded value is _____._____ _____

Decimal Numbers and Percents

o The word percent comes from the Latin *per centum*, meaning per

_____.

o Percent means _____, or the ratio of a number to 100.

o The symbol _____ is called the percent sign. This sign has the same

meaning as the _____ $\dfrac{1}{100}$

o Changing fractions with denominators of 100 to percents:

o Example: $\dfrac{25}{100} = 25\%$

The _____ did not change.

o Example: $\dfrac{3.8}{100} = 3.8\%$

The _____ is not changed, the decimal point doesn't move if

the _____ is 100.

o To change a decimal to a percent:

1. Move the decimal point two places to the _____.

2. Write the _____ sign.

- Example: $0.56 = 56\%$

- Example: $0.345 = 34.5\%$

- Example: $0.02 = 2\%$

o To change percents to a decimal number:

1. Move the decimal two places to the _____.

2. Delete the _____ sign.

- Example: $97\% = 0.97$

- Example: $68.5\% = 0.685$

- Example: $0.64\% = 0.0064$

Fractions and Percents

o If a fraction has denominator _____, it can be changed to a percent by

writing the _____ and adding the _____ sign.

o If the denominator is a factor of _____ (2, 4, 5, 10, 20, 25, or 50), the

fraction can be changed to an equivalent fraction with denominator of

_____ and then changed to a percent.

o When fractions do NOT have factors of 100 as denominators you will need to:

o Change the fraction to decimal form by _____, either by long

division or with a calculator (depending on instructions from instructor).

o Move the decimal two places to the _____ and write the

_____ symbol.

$$\text{Example:} \quad \frac{3}{4} = 4\overline{)3} = 0.75 = 75\%$$

o Helpful calculation tips:

o The numerator goes _____ the division symbol and the

denominator goes _____ of the division symbol.

o If you are using a calculator, you will type the numerator _____,

then the division symbol, and then the _____, followed by

enter/equal.

o To change a percent to a fraction or a mixed number:

o Write the percent as a fraction with 100 as the _____ and drop the

_____ symbol.

○ _____ the fraction, if possible.

Example: $80\% = \dfrac{80}{100} = \dfrac{2 \cdot 2 \cdot 2 \cdot 2 \cdot 5}{2 \cdot 2 \cdot 5 \cdot 5} = \dfrac{4}{5}$

Solving Percent Problems by Using Proportions

o _____ means hundredths and percent is the ratio of a number to 100.

o The percent proportion is:

_____ = _____

o The parts of the percent proportion are:

- P% = _____ (written as the ratio $\dfrac{P}{100}$)

- B = _____ (number that we are finding the percent of)

- A = _____ (a part of the base)

o 25% of 60 is 15 has the following parts:

o Base: _____

o Percent: _____

o Amount: _____

The Cartesian Coordinate System

- The Cartesian coordinate system (or the rectangular coordinate system) is based on a relationship between _____ in a plane and ordered pairs of real numbers.

- For an equation written like $y = 5x + 7$, ordered pairs are written in the form of (__, _____).

- In an ordered pair, x is called the first _____ and y is called the _____ coordinate.

- To find ordered pairs that satisfy an equation in two variables, we can choose any value for one variable and find the corresponding value for the other variable by _____ into the equation.

- There is an _____ number of such ordered pairs. Any real number could have been chosen for x and the corresponding value for y calculated.

- In an ordered pair of the form (x,y), the first coordinate x is called the _____ variable and the second coordinate y is called the _____ variable.

- Label the following information on the coordinate plane below: x-axis, y-axis, quadrants (4), and origin.

- In quadrant I, the x is _____ and the y is _____.

o   In quadrant II, the x is _____ and the y is _____.

o   In quadrant III, the x is _____ and the y is _____.

o   In quadrant IV, the x is _____ and the y is _____.

o   There is a one-to-one _____ between points in a plane and ordered pairs of real numbers.

o   Unless otherwise stated, assume that the grid lines are _____ unit apart.

o   To plot an x-coordinate, you will move_____ for negative values and _____ for positive values.

o   To plot any y-coordinate, you will move_____ for negative values and up for _____values.

Graphing Linear Equations in Two Variables: *Ax + By = C*

o   To find some solutions for an equation like 2-2x = y, you will need to:

1.   choose arbitrary values for _____

2.   find the corresponding values for _____ by substituting into the equation.

o   The standard form of an equation is _____ where A, B, and C are real numbers and A and B are not both equal to _____, is called the standard form of a linear equation.

o   In the standard form _____, A and B may be positive, negative, or 0, but A and B cannot both equal 0.

o   Every line corresponds to some _____ _____, and the graph of every linear equation is a line.

o   Two points determine a _____.

o   Steps to graph a linear equation in two variables:

1.   Locate any _____ points that satisfy the equation. (Choose values for x and y that lead to _____ solutions. Remember that there is an _____ number of choices for either x or y. But, once a value for x or y is chosen, the corresponding value for the other variable is found by substituting into the equation.)

2.   Plot these _____ points on a Cartesian coordinate system.

3. Draw a _____ through these two points. (Note: Every point on that line will satisfy the equation.)

4. To check: Locate a _____ point that satisfies the equation and check to

see that it does indeed lie on the line.

o By letting $x=$_____, you will locate the point on the graph where the line crosses

(or intercepts) the y-axis. This point is called the y-_____ and is of the form (_____,y).

o The x-intercept is the point found by letting $y=$_____ . This is the point where the line crosses (or intercepts) the x-axis and is of the form (x,__).

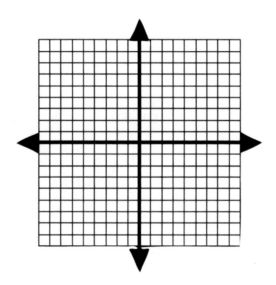

o When the intercepts result in a point with fractional (or decimal) coordinates and

_____ is involved, then a _____ point that satisfies the

equation should be found to verify that the line is graphed correctly.

Exponents I

o   If a is a _____ real number and m and n are integers, then $a^m + a^n = a^{m+n}$.

o   What this means is that in order to multiply two powers with the same base, keep the

_____ and _____ the exponents.

Example: $2^3 \cdot 2^5 = 2^{3+5} = 2^8$

o   If a variable or constant has no exponent written, the exponent is understood to be

_____.

o   Any non-zero value raised to the 0 power is worth _____.

o   The expression 0^0 is _____.

o   To divide two powers with the same base, keep the _____ and _____

the exponents.

o   In other words, subtract the _____ exponent from the _____ _____ exponent.

Example:

$$\frac{d^8}{d^3} = d^{8-3} = d^5$$

o   If a is a nonzero real number and n is an _____, then $a^{-n} = \dfrac{1}{a^n}$.

Example:

$$n^{-2} = \frac{1}{n^2}$$

Greatest Common Factor of Two or More Terms

o The result of multiplication is called the _____.

o The numbers or expressions being multiplied are called _____ of the product.

o The reverse of multiplication is called _____.

o The _____ (GCF) of two or more integers is the largest integer that is a factor (or divisor) of _____ of the integers.

o Steps for Finding the GCF for a set of Terms are:

1. Find the _____ _____ of all integers and integer coefficients.

2. List all the factors that are common to _____ terms, including variables.

3. Choose the _____ power of each factor that is common to all terms.

4. _____ these powers to find the GCF.

o If there is no common prime factor or variable, then the GCF is _____.

Factoring Trinomials with Leading Coefficient 1

o   To factor $x^2 + bx + c$, if possible, find an integer pair of _____ of c whose

_____ is b.

1. If c is _____, then both factors must have the _____ sign.

- Both will be positive if b is _____.

- Both will be negative if b is _____.

2. If c is _____, then one factor must be _____ and the other

_____.

o   One additional thing to keep in mind is when factoring polynomials, always look for a

common _____ factor first.

o   Then, if there is one, remember to include this _____ monomial factor as part

of the answer.

o   If there is a common monomial factor, factor out this common monomial factor

_____ and then factor the remaining _____, if possible.

o   Not all polynomials are _____.

o   For example, no matter what combinations are tried, $x^2 + 3x + 4$ does not have two binomial

factors with integer coefficients. (There are no factors of +4 that will add to +3.) The

polynomial is not factorable (or _____).

o   A polynomial is not _____ if it cannot be factored as the product of polynomials

with integer coefficients.

Factoring Trinomials by Trial and Error

- The ac-method of factoring is in reference to the coefficients a and c in the general

  form $ax^2 + bx + c$ and involves the method of factoring by _____.

- Complete the table below describing each step of the ac-method and an example:

General Method $ax^2 + bx + c$	Example $2x^2 + 9x + 10$
Step 1: Multiply $a \cdot c$.	$2 \cdot 10 = 20$
Step 2: Find two integers whose product is ac and whose sum is b. If this is not possible, then the trinomial is not factorable.	Find two integers whose product is 20 and whose sum is 9.
Step 3: Rewrite the middle term (bx) using the two numbers found in Step 2 as coefficients.	Rewrite the middle term $(+9x)$ using $+4$ and $+5$ as coefficients.
Step 4: Factor by grouping the first two terms and the last two terms.	Factor by grouping the first two terms and the last two terms.
Step 5: Factor out the common binomial factor. This will give two binomial factors of the trinomial $ax^2 + bx + c$.	Factor out the common binomial factor $(x+2)$.

- Helpful hints for factoring:

  - When factoring polynomials, always look for the _____

    _____ _____ first. Then, if there is one, remember
    to include this common factor as part of the _____.

- To factor completely means to find factors of the polynomial such that none of the factors are themselves _____.

- Not all polynomials are _____. Any polynomial that cannot be factored as the product of polynomials with integer coefficients is not factorable.

- Factoring can be checked by _____ the factors. The product should be the original expression.

○ No matter which method you use (the *ac*-method or the trial-and-error method), factoring trinomials takes _____. With practice you will become more efficient with either method. Make sure to be _____ and observant.

Special Factorizations - Squares

o   Remember that the _____ of squares rule is $x^2 - a^2 = (x+a)(x-a)$.

o   When factoring difference of squares:

1.   Take the square root of the _____ term.

2.   Take the square root of the _____ term.

3.   Form the two binomials, one with an _____ symbol and one with a _____ symbol.

o   The sum of two squares is an expression of the form $x^2 + a^2$ and is _____ factorable, unless there is a GCF that can be factored out of _____ terms.

o   Factoring a perfect square trinomial gives the _____ of a binomial.

o   In a perfect square trinomial, both the _____ and _____ terms of the trinomial must be perfect squares.

o   If the first term is of the form x^2 and the last term is of the form a^2, then the middle term must be of the form _____ or _____.

Quadratic Equations: The Quadratic Formula

o   The quadratic formula gives the _____ of any quadratic equation in terms of the coefficients a, b, and c of the general quadratic equation.

o   To develop the quadratic formula, we solve the general quadratic equation by

_____ the _____.

o   The _____ _____ is $x = \dfrac{-b \pm \sqrt{b^2 - 4ac}}{2a}$ when $a \neq 0$.

o   The quadratic formula should be _____.

o   Solve the quadratic equation $7x^2 - 2x + 1 = 0$ by using the quadratic formula:

   Step 1:

   Step 2:

   Step 3:

   Step 4:

   Step 5:

   Step 6:

o   Avoid making a mistake of simplifying fractions by dividing the denominator into only

_____ of the terms in the numerator.

o   The expression $b^2 - 4ac$, the part of the quadratic formula that lies under the radical sign, is called the _____.

o   The discriminant identifies the _____ and _____ of solutions to a quadratic equation.

o   There are three possibilities: the discriminant is either:

1.

2.

3.

o   Complete the table below describing the different types of discriminants and the nature of

solutions:

Discriminant	Nature of Solutions
	Two real solutions
$b^2 - 4ac = 0$	
	Two nonreal solutions

Solving Proportions

o   Steps to solve a proportion:

    o   Find the cross _____ (or cross _____).

    o   _____ both sides of the equation by the coefficient of the variable.

    o   _____.

o   When you are solving a proportion, it will be helpful for you to write each new

    equation below the previous equation. Keep the = signs _____.

    This will help you organize the steps and avoid simple _____.

    Example:

$\dfrac{2}{6} = \dfrac{5}{x}$ Use cross products.

$2x = 30$ Divide both sides by 2.

$\dfrac{2x}{2} = \dfrac{30}{2}$ Simplify.

$x = 15$

o   Steps to solve a word problem with a proportion:

    o   Identify the unknown quantity and use a _____ to represent this quantity.

    o   Set up a _____ in which the units are compared in the _____ order.
    (Make sure that the units are labeled so they can be seen to be in the right order.)

    o   _____ the proportion.

Example:

$\dfrac{25\,miles}{10\,hours} = \dfrac{x\,miles}{45\,hours}$ Use cross products.

$10x = 1125$ Divide both sides by 10.

$\dfrac{10x}{10} = \dfrac{1125}{10}$ Simplify.

$x = 11.25$

Square Roots and the Pythagorean Theorem

o   A number is squared when it is _____ by _____.

o   The result of squaring a number is called a _____ _____.

o   Some terminology for radicals:

   ▪   The symbol _____ is called a radical sign.

   ▪   The number under the radical sign is called the _____.

   ▪   The complete expression, such as _____, is called a radical or radical

      expression.

o   Most square roots are _____ _____ (infinite nonrepeating decimals).

   This means that most square roots can be only approximated with _____.

o   An example of this is _____.

o   Steps to find a square root on a calculator are:

   1.

   2.

   3.

Using your calculator and the three steps listed above, find $\sqrt{17}$ and round to the nearest thousandth. The

$\sqrt{17}$ rounded to the nearest thousandth is _____.

o Label the following parts of the

right triangle on the image to the right:

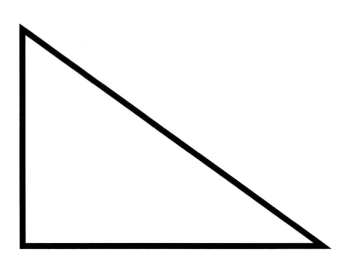

1. 90 degree angle

2. Hypotenuse

3. Leg

o The Pythagorean Theorem states that in a _____ triangle, the square of the length of the _____ (c) is equal to the _____ of the squares of the lengths of the two _____ (a and b).

o The formula for the Pythagorean Theorem is _____.

o Label the parts of the right triangle that are

used in the Pythagorean Theorem on the

image to the right:

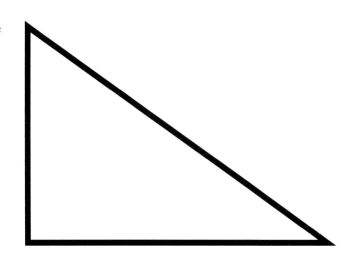

1. Leg a

2. Leg b

3. Hypotenuse c

Simplifying Algebraic Expressions

o   An _____ expression is a combination of variables and numbers using any of the _____ of addition, subtraction, multiplication, or division, as well as _____.

o   Three examples of algebraic expressions are:

1.

2.

3.

o   To combine like terms, _____ or _____ the coefficients and _____ the common variable expression.

Evaluating Algebraic Expressions

- o The expression $-x^2$ will have a _____ value.

    - ▪ An example of this is $-6^2 =$ _____ .

- o The expression $(-x)^2$ will have a _____ value.

    - ▪ An example of this is $(-6)^2 =$ _____ .

- o To evaluate an algebraic expression, you will need to:

    1.

    2.

    3.

Working with Formulas

o   Formulas are general _____ or principles stated _____.

Below is a table of formulas and their meanings.   You will use these in this lesson.

Formula	Meaning
$I=Prt$	The simple interest I earned by investing money is equal to the product of the _____P times the rate of interest r times the time t in one year or part of a year.
$C = \dfrac{5}{9}(F-32)$	Temperature in degrees Celsius C equals $\dfrac{5}{9}$ times the _____between the Fahrenheit temperature F and _____.
$d=rt$	The _____ traveled d equals the product of the rate of _____ r and the time t.
$P=a+b+c$	The perimeter P of a triangle is equal to the sum of the _____ of the three sides a, b, and c.
$F = kAv^2$	The lifting _____ F is equal to the constant k by the area A and by the square of the plane's velocity v.

Multiplication and Division with Fractions and Mixed Numbers

o   To multiply with mixed numbers, you will need to change the mixed numbers to _____

_____ and then multiply the fractions.

o   You can multiply fractions and mixed numbers while reducing at the same time by using

_____ factors.

o   Multiply and reduce to lowest terms:

$$1\frac{2}{5} \cdot \frac{5}{8} = \frac{\qquad\cdot\qquad}{\qquad\cdot\qquad\cdot\qquad} = \underline{\qquad}$$

o   Remember that if all of the factors in the _____ or the _____ divide

out, then the value of _____ must be used as a factor.

o   The reciprocal of $\frac{a}{b}$ is $\frac{b}{a}$ (when a and b do not equal 0).
The product of a nonzero number and its reciprocal is always _____.

o   Match the following values and its reciprocal:

15   2

$\frac{7}{8}$ $\frac{1}{15}$

$\frac{1}{2}$ $\frac{8}{7}$

o   To divide fractions by any nonzero number, you will _____ by its reciprocal.

For example, $\frac{1}{2} \div \frac{2}{5} = \frac{1}{2} \cdot \underline{\quad}$.

Least Common Multiple (LCM)

○ The _____ of a number are the products of that number with the counting numbers.

○ The least common multiple (LCM) of two (or more) whole numbers is the _____ number that is a multiple of each of these numbers.

○ Steps to find the LCM of Counting Numbers:

1. Find the _____ _____ of each number.

2. Identify the prime factors that appear in _____ one of the prime factorizations.

3. Form the product of these primes using each prime the _____ number of times it appears in any one of the prime factorizations.

○ Find the LCM of 24 and 36.

1. The prime factorization for 24 is: _____ .

The prime factorization for 36 is: _____ .

2. You will use the factors of _____.

3. Multiply the factors together to get the LCM of _____ .

Addition and Subtraction with Fractions

o   To **add** fractions with the **same denominator**, you will need to:

 o   _____ the numerators.

 o   Keep the _____ denominator.

 o   _____, if possible.

o   To **add** fractions with **different denominators**, you will need to:

 o   Find the least common _____ (LCD). Remember that the LCD is the least

  common multiple for the denominators.

 o   Change each fraction into an _____ fraction with that denominator.

 o   _____ the new fractions.

 o   Reduce, if possible.

o   A common error that occurs when adding fractions that needs to be avoided is

_____ out across the addition sign.

o   To **subtract** fractions with the **same denominator**, you will need to:

 o   _____ the numerators.

 o   Keep the _____ denominator.

 o   Reduce, if possible.

o   To **subtract** fractions with **different denominators**, you will need to:

 o   Find the least common _____ (LCD).

 o   Change each fraction into an _____ fraction with that denominator.

 o   _____ the new fractions.

 o   Reduce, if possible.

Decimals and Fractions

o   To change from decimal numbers to fractions:

  o   A terminating decimal number can be written in fraction form by writing a fraction with the following:

  ▪   a _____ that consists of the whole number formed by all the digits of the decimal number

  ▪   a _____ that is the power of ten that names the position of the last digit on the right

o   To change from a fraction to a decimal number:

  o   To change a fraction to decimal form, we divide the numerator by the denominator.

  ▪   If the remainder is eventually 0, the decimal number is said to be _____.

  ▪   If the remainder is never 0, the decimal number is said to be _____.

o   Nonterminating decimal numbers can be _____ or nonrepeating.

o   A _____ repeating decimal (also called an infinite repeating decimal number) has a repeating pattern to its digits.

o   Every fraction with a whole number numerator and nonzero denominator is either terminating or repeating. Such numbers are called _____ numbers.

o   Sometimes, changing fractions to decimal form may involve _____ the decimal form of a number and settling for an approximate answer. To have a more accurate answer, we may need to change the number from decimal form to _____ form and then perform the operations.

Decimals and Percents

o   The word percent comes from the Latin *per centum*, meaning per

   _____.

o   Percent means _____, or the ratio of a number to 100.

o   The symbol _____ is called the percent sign. This sign has the same

   meaning as the _____ $\frac{1}{100}$.

o   Changing fractions with denominators of 100 to percents:

o   Example: $\frac{25}{100} = 25\%$

   The _____ did not change.

o   Example: $\frac{3.8}{100} = 3.8\%$

   The _____ is not changed, the decimal point doesn't move if

   the _____ is 100.

o   To change a decimal to a percent:

1.   Move the decimal point two places to the _____.

2.   Write the _____ sign.

   ▪   Example: $0.56 = 56\%$

   ▪   Example: $0.345 = 34.5\%$

   ▪   Example: $0.02 = 2\%$

o   To change percents to a decimal number:

1.   Move the decimal two places to the _____.

2. Delete the _____ sign.

- Example: 97% = 0.97

- Example: 68.5% = 0.685

- Example: 0.64% = 0.0064

Fractions and Percents

o   If a fraction has denominator _____, it can be changed to a percent by

writing the _____ and adding the _____ sign.

o   If the denominator is a factor of _____ (2, 4, 5, 10, 20, 25, or 50), the

fraction can be changed to an equivalent fraction with denominator of

_____ and then changed to a percent.

o   When fractions do NOT have factors of 100 as denominators you will need to:

o   Change the fraction to decimal form by _____, either by long

division or with a calculator (depending on instructions from instructor).

o   Move the decimal two places to the _____ and write the

_____ symbol.

Example: $\dfrac{3}{4} = 4\overline{)3} = 0.75 = 75\%$

o   Helpful calculation tips:

o   The numerator goes _____ the division symbol and the

denominator goes _____ of the division symbol.

o   If you are using a calculator, you will type the numerator _____,

then the division symbol, and then the _____, followed by

enter/equal.

o   To change a percent to a fraction or a mixed number:

o   Write the percent as a fraction with 100 as the _____ and drop the

_____ symbol.

○ _____ the fraction, if possible.

Example: $80\% = \dfrac{80}{100} = \dfrac{2\cdot2\cdot2\cdot2\cdot5}{2\cdot2\cdot5\cdot5} = \dfrac{4}{5}$

Working with Formulas

o   Formulas are general _____ or principles stated _____.

Below is a table of formulas and their meanings. You will use these in this lesson.

Formula	Meaning
$I=Prt$	The simple interest I earned by investing money is equal to the product of the _____ P times the rate of interest r times the time t in one year or part of a year.
$C = \frac{5}{9}(F - 32)$	Temperature in degrees Celsius C equals $\frac{5}{9}$ times the _____ between the Fahrenheit temperature F and _____.
$d=rt$	The _____ traveled d equals the product of the rate of _____ r and the time t.
$P=a+b+c$	The perimeter P of a triangle is equal to the sum of the _____ of the three sides a, b, and c.
$F = kAv^2$	The lifting _____ F is equal to the constant k by the area A and by the square of the plane's velocity v.

The Cartesian Coordinate System

o   The Cartesian coordinate system (or the rectangular coordinate system) is based on a

relationship between _____ in a plane and ordered pairs of real numbers.

o   For an equation written like $y = 5x + 7$, ordered pairs are written in the form of (__, _____).

o   In an ordered pair, x is called the first _____ and y is called the _____

coordinate.

o   To find ordered pairs that satisfy an equation in two variables, we can choose any value for

one variable and find the corresponding value for the other variable by _____

into the equation.

o   There is an _____ number of such ordered pairs. Any real number could have

been chosen for x and the corresponding value for y calculated.

o   In an ordered pair of the form (x, y), the first coordinate x is called the _____

variable and the second coordinate y is called the _____ variable.

o   Label the following information on the coordinate plane below: x-axis, y-axis, quadrants (4),

and origin.

o   In quadrant I, the x is _____ and the y is _____.

o   In quadrant II, the x is _____ and the y is _____.

o   In quadrant III, the x is _____ and the y is _____.

o   In quadrant IV, the x is _____ and the y is _____.

o   There is a one-to-one _____ between points in a plane and ordered pairs of real numbers.

o   Unless otherwise stated, assume that the grid lines are _____ unit apart.

o   To plot an x-coordinate, you will move_____ for negative values and _____ for positive values.

o   To plot any y-coordinate, you will move_____ for negative values and up for _____values.

Graphing Linear Equations in Two Variables: Ax + By = C

- To find some solutions for an equation like $2 - 2x = y$, you will need to:

    1. choose arbitrary values for _____

    2. find the corresponding values for _____ by substituting into the equation.

- The standard form of an equation is _____ where A, B, and C are real numbers and A and B are not both equal to _____, is called the standard form of a linear equation.

- In the standard form _____, A and B may be positive, negative, or 0, but A and B cannot both equal 0.

- Every line corresponds to some _____ _____, and the graph of every linear equation is a line.

- Two points determine a _____.

- Steps to graph a linear equation in two variables:

    1. Locate any _____ points that satisfy the equation. (Choose values for x and y that lead to _____ solutions. Remember that there is an _____ number of choices for either x or y. But, once a value for x or y is chosen, the corresponding value for the other variable is found by substituting into the equation.)

    2. Plot these _____ points on a Cartesian coordinate system.

    3. Draw a _____ through these two points. (Note: Every point on that line will satisfy the equation.)

4. To check: Locate a _____ point that satisfies the equation and check to

see that it does indeed lie on the line.

o   By letting $x =$_____, you will locate the point on the graph where the line crosses

(or intercepts) the y-axis. This point is called the y-_____ and is of the form (__,y).

o   The x-intercept is the point found by letting $y =$_____ .This is the point where the line

crosses (or intercepts) the x-axis and is of the form (x,_____).

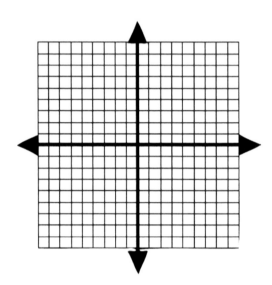

o   When the intercepts result in a point with fractional (or decimal) coordinates and

_____ is involved, then a _____ point that satisfies the equation

should be found to verify that the line is graphed correctly.

o   For real numbers a and b, the graph of $y = b$ is a _____ line and $x = a$ is a

_____ line.

The Slope-Intercept Form: y = mx + b

○ For a line, the ratio of _____ to _____ is called the slope of the line.

○ Plot the rise and run between points on the line in the graph below:

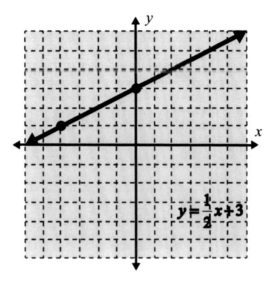

$$y = \frac{1}{2}x + 3$$

○ The concept of slope also relates to situations that involve _____ of

_____.

○ _____ Formula = $m = \dfrac{rise}{run} = \dfrac{y_2 - y_1}{x_2 - x_1}$

○ The variable that represents slope is _____.

○ The slope is the same even if the order of the points is _____.

○ The coordinates must be _____ in the same order in both the _____

and the _____.

○ Lines with positive slope go up (_____) as we move along the line

from _____ to _____.

o   Lines with negative slope go down(_____) as we move along the line

from _____ to _____.

o   If two points have the same _____, such as (2, 8) and (7,8), then the line through

these two points will be _____.

o   If two points have the same _____, such as (2, 3) and (2,−2), then the line through

these two points will be _____.

o   The following two general statements are true for horizontal and vertical lines:

1. For horizontal lines (of the form $y = b$), the slope is _____.

2. For vertical lines (of the form $x = a$), the slope is _____.

o $y=mx+b$ is called the _____-_____ form for the equation of a line,

where m is the _____ and $(0, b)$ is the _____.

Evaluating Radicals

o   Complete the tables of Squares and Perfect Squares below:

Integers (n)	1	2	3	4	5	6	7	8	9	10
Perfect Squares (n^2)										

Integers (n)	11	12	13	14	15	16	17	18	19	20
Perfect Squares (n^2)										

o   Finding the square root of a number is the opposite of _____ a number.

o   The symbol _____ is called a radical sign.

o   The number under the radical sign is called the _____.

o   The complete expression, such as _____ , is called a radical or radical expression.

o   If a is a nonnegative real number, then _____ is the principal square root of a.

o   If a is a nonnegative real number, _____ is the negative square root of a.

o   A number is cubed when it is used as a factor 3 times.

o   Complete the cubes of numbers below:

$1^3 =$	$2^3 =$	$3^3 =$	$4^3 =$	$5^3 =$

$6^3 =$	$7^3 =$	$8^3 =$	$9^3 =$	$10^3 =$

o   If a is a real number, then _____ is the cube root of a.

o   In the cube root expression $\sqrt[3]{a}$, the number 3 is called the _____.

o   In a square root expression such as $\sqrt{a}$, the index is understood to be 2 and is _____ written.

o   Expressions with square roots and cube roots, as well as other roots, are called
_____ .

Reading and Writing Whole Numbers

o   The symbols used in our number system are called _____.

o   The value of a digit depends on its position to the left of a beginning point, called a

_____ point.

o   Fill in the missing items in the table below.

1,000,000,000	Billions
100,000,000	
10,000,000	Ten millions
	Millions
100,000	Hundred thousands
10,000	
	Thousands
100	Hundreds
10	
1	Ones

o   The _____ numbers are the number 0 and the natural numbers.

o   _____ notation can help you to translate a whole number into its English word equivalent.

o   Example:

7,034,165 = 7,000,000 + 30,000 + 4000 + 100 + 60 + 5

7,034,165 is read as

Seven _____ , _____ thousand, one hundred sixty-five

Addition and Subtraction with Whole Numbers

- The numbers being added are called _____, and the result of the addition is called the _____.

- In the addition example below, name each term.

$$
\begin{array}{r}
4 \\
+\ 8 \\
\hline
12
\end{array}
$$

4   ←   _____

+ 8   ←   _____

12   ←   _____

- If the sum of the digits in one column is more than 9, you need to _____ the _____ digit of the sum as a number to be added to the next column to the left.

- A _____ is a symbol (generally a letter of the alphabet) that is used to represent an unknown number or any one of several numbers.

- The Properties of Addition:

  - By the _____ Property of Addition, for any whole numbers a and b, $a + b = b + a$.

  - By the Associative Property of Addition, for any whole numbers a, b, and c, $(a + b) + c =$ _____ .

  - By the Additive Identity Property, for any whole number a, $a + \underline{\ \ \ } = a$.

- The _____ of a geometric figure is the distance around the figure.

o _____ is reverse addition. To subtract, we must know how to add. The

missing addend is called the _____.

o In the subtraction example below, name each term.

$$
\begin{array}{r}
12 \\
-\ 8 \\
\hline
4
\end{array}
$$

12 ← _____

− 8 ← _____

4 ← _____

Exponents and Order of Operations

o   When looking at 5^3, you have the following parts:

- The base is: _____

- The exponent is: _____

- The product is : _____

- The exponential expression is: _____

o   In expressions with exponent 2, the base is said to be _____.

o   In expressions with exponent 3, the base is said to be _____.

o   With other exponents, the base is said to be "_____."

o   For any number, a, $a^1 = a$. An example would be $7^1 =$_____.

o   For any nonzero number, a, $a^0 = 0$. An example would be $7^0 =$ _____.

o   The rules for order of operations are:

1.   Simplify within _____ _____, such as

     parentheses ( ), brackets [ ], or braces { }. Start with the _____ group.

2.   Evaluate any numbers or expressions indicated by _____.

3.   Moving from _____ to _____, perform any _____

     or _____ in the order in which it appears.

4.   Moving from _____ to _____, perform any _____

     or _____ in the order in which it appears.

Introduction to Decimal Numbers

o   Decimal notation uses a _____ _____ system and a

_____ point, with whole numbers written to the

_____ and fractions written to the _____ of the

decimal point.

o   To read or write a decimal number:

      o   _____ (or write) the whole number.

      o   Read (or write) the word "_____" in place of the decimal point.

      o   Read (or write) the _____ part as a whole number. Then name the fraction

      with the name of the last _____ on the right. Add "th" to the end of the

      fraction place value.

      o   Remember that if there is not a whole number, you can put a _____ to the

      left of the decimal point.

o   Write the following mixed number as a decimal number:

$$3\frac{4}{100} = \underline{\qquad}.\underline{\qquad}\,\underline{\qquad}$$

and it would be read as _____ AND _____ _____

o   When writing seventeen and 5 thousandths in decimal notation you would have

_____ holders in the tenths and hundredths place values.

It would look like 17.005.

o   When comparing decimals:

      o   Moving _____ to _____, compare digits with the

      same _____ value.

o   When one compared digit is larger, the _____ is larger.

o   Compare the following values:

5.789   Notice that the numbers are lined up, for easier comparison.
5.754

When moving from left to right, the digits are the same until the _____ place

values. Those are going to be used to compare. Since the 8 is a larger value than 5,

_____ is a larger value.

o   Rules for rounding decimals:

o   Look at the digit to the _____ of the place of desired accuracy.

o   If this digit is 5 or _____, make the digit in the desired place of accuracy

one larger and replace all digits to the right with zeros.

All digits to the left remain unchanged unless a 9 is made one larger. This effectively changes

the 9 to 10 which means the next digit to the left must be increased by 1.

o   If this digit is _____ than 5, leave the digit in the desired place of accuracy

as it is and replace all digits to the right with zeros.

All digits to the left remain unchanged.

o   Zeros **to the right of the place of accuracy** and to the right of the decimal point must be

_____. In this way the place of accuracy is clearly understood. If a

rounded number has a 0 in the desired place of accuracy, then that 0 remains.

Round 13.2687 to the nearest hundredth.

The digit in the hundredths is _____ .

The digit in the place value to the right of the hundredths is _____ .

Since that digit is greater than 5, the 6 in the hundredths place value changes to a 7.

The rounded value is _____._____ _____

Decimals and Percents

o   The word percent comes from the Latin *per centum*, meaning per

_____ .

o   Percent means _____ , or the ratio of a number to 100.

o   The symbol _____ is called the percent sign. This sign has the same

meaning as the _____ $\dfrac{1}{100}$.

o   Changing fractions with denominators of 100 to percents:

o   Example: $\dfrac{25}{100} = 25\%$

The _____ did not change.

o   Example: $\dfrac{3.8}{100} = 3.8\%$

The _____ is not changed, the decimal point doesn't move if

the _____ is 100.

o   To change a decimal to a percent:

1.   Move the decimal point two places to the _____ .

2.   Write the _____ sign.

- Example: $0.56 = 56\%$

- Example: $0.345 = 34.5\%$

- Example: $0.02 = 2\%$

o   To change percents to a decimal number:

1.   Move the decimal two places to the _____ .

2. Delete the _____ sign.

- Example: 97% = 0.97

- Example: 68.5% = 0.685

- Example: 0.64% = 0.0064

Solving Percent Problems by Using the Equation: $R \cdot B = A$

- o   Basic equation for solving percent problems is $rate \cdot base = amount$ or $R \cdot B = A$

- o $R =$ _____, always written in _____ form

- o $B =$ _____, the number we are finding the

    _____ of

- o $A =$ _____, part of the _____

    - ▪   _____ means to multiply $(\cdot)$

    - ▪   _____ means equals $(=)$

- o   There are _____ basic types of percent problems using $R \cdot B = A$

- o   Type 1: Find the amount given the _____ and the percent.

    Example: What is 75% of 600? $A = 0.75 \cdot 600$

- o   Type 2: Find the base given the _____ and the amount.

    Example: 15% of what number is 157.5? $0.15 \cdot B = 157.5$

- o   Type 3: Find the percent given the base and the _____.

    Example: What percent of 87 is 115? $R \cdot 87 = 115$

- o   The operations in these examples can be performed with a _____

    or by _____.

- o   Solving it either way, the equations should be written so that the

    _____ signs are aligned one under the other.

- o   Also, writing the equations and the calculated values will help you remember

    whether you are _____ or _____.

Evaluating Algebraic Expressions

- o The expression $-x^2$ will have a _____ value.

    - An example of this is $-6^2 = $ _____ .

- o The expression $(-x)^2$ will have a _____ value.

    - An example of this is $(-6)^2 = $ _____ .

- o To evaluate an algebraic expression, you will need to:

    1.

    2.

    3.

Addition and Subtraction With Whole Numbers

o The numbers being added are called _____, and the result of the addition is called the _____.

o In the addition example below, name each term.

$$4 \leftarrow \underline{\hspace{3cm}}$$
$$\underline{+\ 8} \leftarrow \underline{\hspace{3cm}}$$
$$12 \leftarrow \underline{\hspace{3cm}}$$

o If the sum of the digits in one column is more than 9, you need to _____ the _____ digit of the sum as a number to be added to the next column to the left.

o A _____ is a symbol (generally a letter of the alphabet) that is used to represent an unknown number or any one of several numbers.

o The Properties of Addition:

   ▪ By the _____ Property of Addition, for any whole numbers a and b,

   $a + b = b + a$.

   ▪ By the Associative Property of Addition, for any whole numbers a, b, and c,

   $(a + b) + c = \underline{\hspace{2cm}}$.

   ▪ By the Additive Identity Property, for any whole number a, $a + \underline{\hspace{1cm}} = a$.

o The _____ of a geometric figure is the distance around the figure.

o _____ is reverse addition. To subtract, we must know how to add. The

   missing addend is called the _____.

o In the subtraction example below, name each term.

$$
\begin{array}{r}
12 \\
-\ 8 \\
\hline
4
\end{array}
$$

12 ← _____

− 8 ← _____

4 ← _____

Introduction to Decimal Numbers

○ Decimal notation uses a _____ _____ system and a

_____ point, with whole numbers written to the

_____ and fractions written to the _____ of the

decimal point.

○ To read or write a decimal number:

   ○ _____ (or write) the whole number.

   ○ Read (or write) the word "_____" in place of the decimal point.

   ○ Read (or write) the _____ part as a whole number. Then name the fraction

      with the name of the last _____ on the right. Add "th" to the end of the

      fraction place value.

   ○ Remember that if there is not a whole number, you can put a _____ to the left

      of the decimal point.

○ Write the following mixed number as a decimal number:

$$3\frac{4}{100} = \underline{}.\underline{}\ \underline{}$$

   and it would be read as _____ AND _____ _____

○ When writing seventeen and 5 thousandths in decimal notation you would have

_____ holders in the tenths and hundredths place values.

It would look like 17.005.

○ When comparing decimals:

   ○ Moving _____ to _____, compare digits with the same

      _____ value.

   ○ When one compared digit is larger, the _____ is larger.

o   Compare the following values:

5.789   Notice that the numbers are lined up, for easier comparison.
5.754

When moving from left to right, the digits are the same until the _____ place

values. Those are going to be used to compare. Since the 8 is a larger value than 5,

_____ is a larger value.

o   Rules for rounding decimals:

    o   Look at the digit to the _____ of the place of desired accuracy.

    o   If this digit is 5 or _____, make the digit in the desired place of accuracy one

    larger and replace all digits to the right with zeros.

All digits to the left remain unchanged unless a 9 is made one larger. This effectively changes

the 9 to 10 which means the next digit to the left must be increased by 1.

    o   If this digit is _____ than 5, leave the digit in the desired place of accuracy as

    it is and replace all digits to the right with zeros.

All digits to the left remain unchanged.

    o   Zeros **to the right of the place of accuracy** and to the right of the decimal point must be

    _____. In this way the place of accuracy is clearly understood. If a rounded

    number has a 0 in the desired place of accuracy, then that 0 remains.

Round 13.2687 to the nearest hundredth.

The digit in the hundredths is _____ .

The digit in the place value to the right of the hundredths is _____ .

Since that digit is greater than 5, the 6 in the hundredths place value changes to a 7.

The rounded value is _____._____ _____

Decimals and Fractions

o   To change from decimal numbers to fractions:

    o   A terminating decimal number can be written in fraction form by writing a fraction with the following:

        ▪   a _____ that consists of the whole number formed by all the digits of the decimal number

        ▪   a _____ that is the power of ten that names the position of the last digit on the right

o   To change from a fraction to a decimal number:

    o   To change a fraction to decimal form, we divide the numerator by the denominator.

        ▪   If the remainder is eventually 0, the decimal number is said to be _____.

        ▪   If the remainder is never 0, the decimal number is said to be _____.

o   Nonterminating decimal numbers can be _____ or nonrepeating.

o   A _____ repeating decimal (also called an infinite repeating decimal number) has a repeating pattern to its digits.

o   Every fraction with a whole number numerator and nonzero denominator is either terminating or repeating. Such numbers are called _____ numbers.

o   Sometimes, changing fractions to decimal form may involve _____ the decimal form of a number and settling for an approximate answer. To have a more accurate answer, we may need to change the number from decimal form to _____ form and then perform the operations.

Ratios and Proportions

o   Two meanings for fractions are:

   o   To indicate _____ of a whole.

   o   To indicate _____ .

o   A ratio is a _____ of two quantities by

   _____ .

o   A ratio, 3 days to 4 nights, can be written in three ways:

   1.

   2.

   3.

o   Ratios have the following characteristics:

   o   Ratios can be _____ .

   o   Whenever units of the numbers in a ratio are the same, the ratio has

      _____ units, otherwise known as an _____ number.

   o   When the numbers in a ratio have different units, then the numbers must be

      _____ to clarify what is being compared.

o   A _____ is a statement that two ratios are equal.

o   A proportion is _____ if the cross products are equal.

Angles

o Complete the table below. You will need to write a definition/description for each term, as well as a visual example of the term.

Term	Definition/Description	Visual Example
Point		
Line		
Plane		
Ray		
Angle		
Vertex		
Sides of an angle		
Acute angle		

Right angle		
Obtuse angle		
Straight angle		
Complementary angles		
Supplementary angles		
Congruent angles		
Vertical angles		
Adjacent angles		
Parallel lines		
Perpendicular lines		

Corresponding angles		
Alternate interior angles		

o   Three common ways of labeling angles are:

1.

- Example:

2.

- Example:

3.

- Example

Exponents I

- If a is a _____ real number and m and n are integers, then $a^m + a^n = a^{m+n}$.

- What this means is that in order to multiply two powers with the same base, keep the

  _____ and _____ the exponents.

  Example: $2^3 \cdot 2^5 = 2^{3+5} = 2^8$

- If a variable or constant has no exponent written, the exponent is understood to be

  _____.

- Any non-zero value raised to the 0 power is worth _____.

- The expression 0^0 is _____.

- To divide two powers with the same base, keep the _____ and _____

  the exponents.

- In other words, subtract the _____ exponent from the _____
  _____exponent.

  Example:

  $$\frac{d^8}{d^3} = d^{8-3} = d^5$$

- If a is a nonzero real number and n is an _____, then $a^{-n} = \dfrac{1}{a^n}$.

  Example:

  $$n^{-2} = \frac{1}{n^2}$$

Rationalizing Denominators

o   Calculations of sums and differences are much easier if the denominators are

_____ expressions.

o   The objective is to find an equivalent fraction that has a rational number or an expression

with no _____ for a _____.

o   How to rationalize a denominator containing a square root or cube root:

1.   If the denominator contains a square root, multiply both the numerator and

denominator by an expression that will give a _____ with no

_____ _____.

2.   If the denominator contains a cube root, multiply both the numerator and denominator

by an expression that will give a denominator with no _____ _____

_____.

o   What to if you need to rationalize a denominator containing a sum or difference involving

square roots:

▪   If the denominator of a fraction contains a sum or difference involving a square root,

rationalize the denominator by multiplying both the numerator and denominator by

the _____ of the denominator. This is discussed below:

1. If the denominator is of the form $a-b$, multiply both the numerator and denominator

by _____ .

2. If the denominator is of the form $a+b$, multiply both the numerator and denominator

by _____ .

- The new denominator will be the _____ of two squares and therefore will not contain a radical term.

Quadratic Equations: The Quadratic Formula

- The quadratic formula gives the _____ of any quadratic equation in terms of the coefficients a, b, and c of the general quadratic equation.

- To develop the quadratic formula, we solve the general quadratic equation by

    _____ the _____ .

- The _____ _____ is $x = \dfrac{-b \pm \sqrt{b^2 - 4ac}}{2a}$ when $a \neq 0$.

- The quadratic formula should be _____ .

- Solve the quadratic equation $7x^2 - 2x + 1 = 0$ by using the quadratic formula:

    Step 1:

    Step 2:

    Step 3:

    Step 4:

    Step 5:

    Step 6:

- Avoid making a mistake of simplifying fractions by dividing the denominator into only

    _____ of the terms in the numerator.

- The expression $b^2 - 4ac$, the part of the quadratic formula that lies under the radical sign, is called the _____ .

- The discriminant identifies the _____ and _____ of solutions to a quadratic equation.

o   There are three possibilities: the discriminant is either:

1.

2.

3.

o   Complete the table below describing the different types of discriminants and the nature of

solutions:

Discriminant	Nature of Solutions
	Two real solutions
$b^2 - 4ac = 0$	
	Two nonreal solutions

Exponents and Order of Operations

○ When looking at 5^3, you have the following parts:

- ■ The base is: _____

- ■ The exponent is: _____

- ■ The product is : _____

- ■ The exponential expression is: _____

○ In expressions with exponent 2, the base is said to be _____.

○ In expressions with exponent 3, the base is said to be _____.

○ With other exponents, the base is said to be "_____."

○ For any number, a, $a^1 = a$. An example would be $7^1 =$_____.

○ For any nonzero number, a, $a^0 = 0$. An example would be $7^0 =$ _____.

○ The rules for order of operations are:

1. Simplify within _____ _____, such as

parentheses ( ), brackets [ ], or braces { }. Start with the _____ group.

2. Evaluate any numbers or expressions indicated by _____.

3. Moving from _____ **to** _____, perform any _____

or _____ in the order in which it appears.

4. Moving from _____ **to** _____, perform any _____

or _____ in the order in which it appears.

Ratios and Proportions

o   Two meanings for fractions are:

   o   To indicate _____ of a whole.

   o   To indicate _____.

o   A ratio is a _____ of two quantities by

    _____.

o   A ratio, 3 days to 4 nights, can be written in three ways:

    1.

    2.

    3.

o   Ratios have the following characteristics:

   o   Ratios can be _____.

   o   Whenever units of the numbers in a ratio are the same, the ratio has

    _____ units, otherwise known as an _____ number.

   o   When the numbers in a ratio have different units, then the numbers must be

    _____ to clarify what is being compared.

o   A _____ is a statement that two ratios are equal.

o   A proportion is _____ if the cross products are equal.

Evaluating Algebraic Expressions

o The expression $-x^2$ will have a _____ value.

- An example of this is $-6^2 =$ _____ .

o The expression $(-x)^2$ will have a _____ value.

- An example of this is $(-6)^2 =$ _____ .

o To evaluate an algebraic expression, you will need to:

1.

2.

3.

U.S. Measurements

Fill in the table based on the information given in the Learn portion of this lesson:

U.S. Units of Length	
12 inches (in.) = 1 _____ ( )	3 feet = 1 _____ ( )
36 inches = 1 _____	5280 feet = 1 _____ ( )

U.S. Units of Weight	
16 ounces = 1 _____ ( )	2000 pounds = 1 _____ ( )

U.S. Units of Capacity	
8 fluid ounces ( ) = 1 _____ ( )	2 pints = 1 _____ ( )
2 cups = 1 _____ ( )	4 quarts = 1 _____ ( )
1 pint = 16 _____ _____ ( )	

Units of Time	
60 seconds ( ) = 1 _____ ( )	24 hours = 1 _____
60 minutes = 1 _____ ( )	7 days = 1 _____

○ To convert from one unit of measure, you will either:

    ○ _____ to convert to smaller units. (There will be more smaller units.)

    ○ _____ to convert to larger units. (There will be fewer larger units.)

Metric System: Length & Area

o   The _____ is the basic unit of length in the metric system.

o   Complete the table below by filling in metric unit of measurement information:

Unit	Abbreviation	Amount in Meters
1 millimeter		
		0.01 meter
1 decimeter		0.1 meter
1 meter		
		100 meters
1 kilometer		1000 meters

o   Other common conversions are:

   ▪   1m = _____mm

   ▪   1m = _____ cm

   ▪   1m = _____dm

o   To change to a measure of length from larger to smaller units of measurement, you will:

   ▪   _____ by 10 if the value is one unit smaller.

   ▪   Multiply by _____ if the value is two units smaller.

   ▪   Multiply by _____ if the value is three units smaller.

o   What pattern do you see forming?

_____

o   To change to a measure of length from smaller to larger units of measurement, you will:

▪   _____ by 10 if the value is one unit larger.

▪   Divide by _____ if the value is two units larger.

▪   Divide by _____ if the value is three units larger.

o   What pattern do you see forming?

_____

Label the diagram below to help you with converting metric units of length:

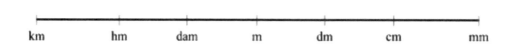

o   In the metric system, a _____ is written to the left of the decimal point if there is no whole number part. An example of this would be _____.

o   In the metric system, no commas are used in writing numbers. If a number has more than _____ digits (left or right of the decimal point), the digits are grouped in _____ from the decimal point with a space between the groups. An example of this would be _____.

U.S. to Metric Conversions

The U.S. customary measure is in degrees _____ (°   ).

The metric measure is in degrees _____ (°   ).

The formula for converting Fahrenheit to _____ is $C = \dfrac{5(F-32)}{9}$.

The formula for converting Celsius to _____ is $F = \dfrac{9C}{5} + 32$.

Length Equivalents (rounded values)	
U.S. to Metric	Metric to U.S.
1 in. = 2.54 _____	1 cm = 0.394 _____
1 ft = 0.305 _____	1 m = 3.28 _____
1 yd = 0.914 _____	1 m = 1.09 _____
1 mi = 1.61 _____	1 km = 0.62 _____

Area Equivalents (rounded values)	
U.S. to Metric	Metric to U.S.
$1 \text{ in.}^2 = 6.45$ _____	$1 \text{ cm}^2 = 0.155$ _____
$1 \text{ ft}^2 = 0.093$ _____	$1 \text{ m}^2 = 10.764$ _____
$1 \text{ yd}^2 = 0.836$ _____	$1 \text{ m}^2 = 1.196$ _____
1 acre = 0.405 _____	1 ha = 2.47 _____

Volume Equivalents (rounded values)	
U.S. to Metric	Metric to U.S.
$1 \text{ in.}^3 = 16.387$ _____	$1 \text{ cm}^3 = 0.06$ _____
$1 \text{ ft}^3 = 0.028$ _____	$1 \text{ m}^3 = 35.315$ _____
1 qt = 0.946 _____	1 L = 1.06 _____
1 gal = 3.785 _____	1 L = 0.264 _____

Mass Equivalents (rounded values)	
U.S. to Metric	Metric to U.S.
1 oz = 28.35 _____	1 g = 0.035 _____
1 lb = 0.454 _____	1 kg = 2.205 _____

Solving Linear Equations: *ax+b=c*

o The general procedure for solving linear equations is now a _____ of the procedures stated in another section.

o Steps for solving equations in the ax+b=c format.

1. _____ like terms on both sides of the equation.

2. Use the _____ principle of equality and add the opposite of the constant *b* to both sides.

3. Use the _____ (or division) principle of equality to multiply both sides by the reciprocal of the coefficient of the variable (or divide both sides by the coefficient itself). The coefficient of the variable will become +1.

4. Check your answer by _____ it for the variable in the original equation.

o Keep in mind that checking can be quite _____-_____ and need not be done for every problem. This is particularly important on _____. You should check only if you have time after the entire exam is _____.

Example:

$3x - 8 = 28$ Add 8 to each side.

$3x = 36$ Divide both sides by 3.

$x = 12$

The Real Number Line and Inequalities

○ A number line is a picture of different types of _____ and their

relationships to each other.

○ The graph of a number is the point that _____ to the number and the

number is called the _____ of the point.

○ On a horizontal number line, the point one unit to the left of 0 is

the _____ of 1. It is called negative 1 and is symbolized −1.

○ The negative sign (−) indicates the _____ of a number as well as a

_____ number. It is also used to indicate subtraction.

○ Take a look at your calculator. The subtraction key on the right side and the negative

key (−) at the bottom of the key pad.

○ The set of numbers consisting of the _____ numbers and their

_____ is called the set of integers: $\mathbb{Z}=\{\ldots,-3,-2,-1,0,1,2,3,\ldots\}$.

○ The _____ numbers are also called _____ integers.

Their opposites are called _____ integers. Zero is its

_____ opposite and is neither positive nor negative ($0 = -0$).

○ The opposite of a positive integer is a _____ integer, and the opposite
of a negative integer is a _____ integer.

○ A rational number is a number that can be written in the form of $\dfrac{a}{b}$, where a and b are

_____ and $b \neq$ _____ . ($\neq$ is read "is not equal to").

○ A rational number is a number that can be written in _____ form as a

_____ decimal or as an infinite _____ decimal.

o   Irrational numbers can be written as infinite _____ decimal numbers.

o   All rational numbers and irrational numbers are classified as _____

    numbers ($\mathbb{R}$) and can be written in some decimal form.

o   Symbols of inequalities:

o   = is _____ to

o   ≠ is _____ equal to

o   < is _____ than

o   > is _____ than

o   ≤ is less than or _____ to

o   ≥ is greater than or _____ to

Multiplication with Whole Numbers

o The result of multiplication is called a _____. The two numbers being

multiplied are called _____ of the product.

o Make note of three symbols that can be used to indicate multiplication. Then write

"4 times 5" using that notation.

Symbol	Write "4 times 5" using the symbol

o The Properties of Multiplication:

- By the Commutative Property of Multiplication, for any whole numbers a and b,

  $a \cdot b =$ _____.

- By the _____ Property of Multiplication, for any whole numbers a, b,

  and c, $(a \cdot b) \cdot c = a \cdot (b \cdot c)$.

- By the Multiplicative Identity Property, for any whole number a, $a \cdot$ ____ $= a$.

- By the Multiplication Property of ____, for any whole number a, $a \cdot$ ____ $= 0$.

o According to the _____ Property of Multiplication over Addition, for any

   whole numbers a, b, and c, $a(b+c) = a \cdot b + a \cdot c$.

o Add arrows to show how a gets multiplied with both b and c. $a(b+c) = a \cdot b + a \cdot c$

o The _____ of a rectangle (measured in square units) is found by multiplying its

   length by its width.

o Determine the area of the rectangle below by finding the length and width. Assume each

   interior square measures 1 inch by 1 inch.

Length: _____ inches   Width: _____ inches   Area: _____ square inches

Division with Whole Numbers

○ In the division example below, name each term.

_____

↓

$$\frac{12}{4} = 3 \quad \leftarrow \quad \underline{\hspace{3cm}}$$

↑

_____

○ The number left after division is called the _____.

○ Factors are also called _____ because each factor will _____

evenly into the product.

○ For any number a, $\dfrac{a}{1} = \underline{\hspace{1cm}}$.

○ For any nonzero number a, $\dfrac{a}{a} = \underline{\hspace{1cm}}$.

○ Division Involving 0:

  ▪ If a is any nonzero whole number, then $\dfrac{0}{a} = \underline{\hspace{1cm}}$.

  ▪ If a is any whole number, then $\dfrac{a}{0}$ is _____ .

o   In the division example below, name each term.

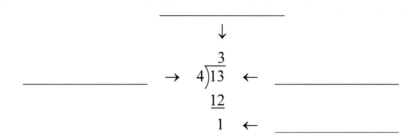

o   Remember that the remainder must be _____ than the divisor.

o   If the remainder is 0, we say that both factors _____ _____

into the dividend.

Tests for Divisibility

o   If a number can be divided by another number so that the _____ is 0, then the _____ is exactly divisible by (or is divisible by) the _____ _____. In other words, the divisor divides the dividend.

o   Even whole numbers are divisible by _____.

o   If a whole number is divided by 2 and the remainder is 0 then the whole number is _____.

o   Odd whole numbers are _____ divisible by 2.

o   If a whole number is divided by 2 and the remainder is 1, then the whole number is _____.

o   Every whole number is either _____ or _____.

o   Divisibility rules are:

   ▪   If the last digit (ones digit) of a whole number is 0, 2, 4, 6, or 8 (an even digit), then the number is divisible by _____ (the number is even).

   ▪   If the sum of the digits of a whole number is divisible by 3, then the number is divisible by _____.

   ▪   If the _____ two digits of a whole number form a number that is divisible by 4, then the number is divisible by _____.

   ▪   If the last digit (ones digit) of a whole number is _____ or _____, then the number is divisible by 5.

   ▪   If a whole number is divisible by both _____ and 3, then the number is divisible by 6.

   ▪   If the sum of the digits of a whole number is divisible by 9, then the number is divisible by _____.

   ▪   If the _____ digit (ones digit) of a whole number is 0, then the number is divisible by 10.

Exponents I

o If a is a _____ real number and m and n are integers, then $a^m + a^n = a^{m+n}$.

o What this means is that in order to multiply two powers with the same base, keep the

   _____ and _____ the exponents.

   Example: $2^3 \cdot 2^5 = 2^{3+5} = 2^8$

o If a variable or constant has no exponent written, the exponent is understood to be

   _____.

o Any non-zero value raised to the 0 power is worth _____.

o The expression 0^0 is _____.

o To divide two powers with the same base, keep the _____ and _____

   the exponents.

o In other words, subtract the _____ exponent from the _____
   _____exponent.

   Example:

   $$\frac{d^8}{d^3} = d^{8-3} = d^5$$

o If a is a nonzero real number and n is an _____, then $a^{-n} = \dfrac{1}{a^n}$.

   Example:

   $$n^{-2} = \frac{1}{n^2}$$

Exponents II

○ If a is a nonzero real number and m and n are integers, then $\left(a^m\right)^n =$ _____ . In other words,

the value of a power raised to a power can be found by _____ the exponents and

keeping the base.

○ If a and b are nonzero real numbers and n is an integer, then $\left(ab\right)^n =$ _____ . In words, a

power of a product is found by raising each _____ to that power.

○ Be careful with the location of a negative sign. Is -7^2 equal to $\left(-7\right)^2$?

○ If a and b are nonzero real numbers and n is an integer, then $\left(\dfrac{a}{b}\right)^n =$ _____ . In words, a

power of a quotient (in fraction form) is found by raising both the _____ and the

_____ to that power.

○ When dealing with fractions involving negative exponents, $\left(\dfrac{a}{b}\right)^{-n} =$ _____ .

Evaluating Radicals

o   Complete the tables of Squares and Perfect Squares below:

Integers (n)	1	2	3	4	5	6	7	8	9	10
Perfect Squares (n^2)										

Integers (n)	11	12	13	14	15	16	17	18	19	20
Perfect Squares (n^2)										

o   Finding the square root of a number is the opposite of _____ a number.

o   The symbol _____ is called a radical sign.

o   The number under the radical sign is called the _____.

o   The complete expression, such as _____ , is called a radical or radical expression.

o   If a is a nonnegative real number, then _____ is the principal square root of a.

o   If a is a nonnegative real number, _____ is the negative square root of a.

o   A number is cubed when it is used as a factor 3 times.

o   Complete the cubes of numbers below:

$1^3 =$	$2^3 =$	$3^3 =$	$4^3 =$	$5^3 =$

$6^3 =$	$7^3 =$	$8^3 =$	$9^3 =$	$10^3 =$

o   If a is a real number, then _____ is the cube root of a.

o   In the cube root expression $\sqrt[3]{a}$, the number 3 is called the _____.

o   In a square root expression such as $\sqrt{a}$, the index is understood to be 2 and is _____ written.

o   Expressions with square roots and cube roots, as well as other roots, are called _____ .